HEARTS OF GOLD

Reflections of

STEM

Gold Award Girl Scouts

Sheryl M Robinson

Copyright Page

Published by Grow and Share Network, LLC
First Edition, 2026

ISBN: 978-1-972135-00-6

Printed in the United States of America

Books in the Hearts of Gold Series

- *Earth Guardian*
- *STEM*
- *Creative Voice*
- *Health*
- *Inclusion*
- *Advocacy*
- *Community Connector*

Table of Contents

Chapter 1
STEM Superpowers

The Spark Of A Social Solution

If you look around your neighborhood, you might notice things that seem small at first: a long line at a fast-food window, a classmate who doesn't have a computer at home, or a local landmark that no one seems to know the history of. But to a STEM innovator, these moments aren't small at all—they are signals. They are clues about what a community needs and how you can use science, technology, engineering, and math to help. Imagine interning at a medical clinic and watching patients receive life-saving advice to "eat healthier," only to see them walk right into a nearby fast-food restaurant because fresh food is too expensive or too far away. This is the reality of a "food desert," where accessibility and practical knowledge are the actual barriers to health. Instead of just handing out pamphlets—which are often unsustainable and ignored—an innovator uses data to bridge the gap. By identifying neighborhoods without grocery stores and holding interactive sessions, you can demonstrate that healthy, nutritious meals can be prepared for under $10. This simple comparison empowers families to choose wellness over a typical fast-food burrito.

STEM is a creative superpower that allows you to identify a "fundamental deficiency" and provide the nourishment a community needs to bloom. It might mean looking halfway across the world to a remote village struggling with a serious "education divide". You might see children who are bright and eager but trapped in a cycle of poverty because they are isolated from urban resources and digital tools. Providing a hand, or "one push," through a permanent computer lab can give these students the same resources found in first-world countries. These solutions prove that you don't have to wait to grow up to make a difference; you can start right now. Every big project begins with one small choice and a moment of curiosity. Whether it is recognizing a need for better nutrition or a lack of technology, STEM is the toolkit that helps you turn that care into a concrete solution. By paying attention to the clues around you and believing in your power to solve them, you are already walking the path of an innovator.

Engineering Change Across Borders
⊕ ⊕ ⊕ ⊕ ⊕

STEM allows young protectors to look to global industries for creative solutions. Imagine traveling to a neighborhood marked by poverty and meeting women who create recycled paper products but must use old deodorant containers to smooth the sheets. This manual technique is "very time-consuming and hard," but it provides an opportunity for an engineering intervention. By stepping into the "realm" of engineering to design a custom stainless-steel tool, you can streamline their work and improve their livelihoods. Stepping outside of your comfort zone creates a "ripple effect" that changes lives thousands of miles away. STEM also provides a way to tackle the "gender gap" in fields like robotics. By making science accessible to underrepresented communities, an innovator can transform from a student into a role model who mentors international teams.

This global vision ensures that children everywhere have the resources to succeed, even if they currently feel like the "only one in the room". One project involved launching a robotics program in a low-income school that tripled in size within three years, eventually becoming completely self-sustaining. Others use technology to solve local mysteries, like building a location-aware walking tour app to share "hidden gems" of history that are otherwise inaccessible due to limited tours or hilly terrain. These innovators aren't just completing a project; they are restoring damaged systems and redesigning how people interact with their world. You might find yourself coordinating guest speakers from professional pilot organizations or working with local airports to secure facilities for workshops. Whether through metal fabrication, digital code, or aviation, these stories prove that ordinary kids can spot what doesn't feel right and change it. If you have the passion to see it through, you are the only person who sets your own limits.

Chapter 2
Girls in Aviation

Autumn Pepper Rhodes (Ep 151)

Autumn Pepper Rhodes' journey was fueled by the powerful feeling she experienced at just ten years old: the rush of her "love at first flight". It was during a flight provided through the Experimental Aircraft Association (EAA), a program that opened her eyes to the thrilling world of aviation. Once that spark was ignited, Pepper knew she wanted to pursue flying, eventually joining organizations like the Civil Air Patrol.

However, as she continued to pursue her passion, a clear problem emerged. The aviation programs she joined were "predominantly male" and often consisted of older gentlemen, leaving her without the kind of female mentors she needed as a young pilot. She recognized that if she had this gap in her own experience, other young women must be missing out, too.

This realization became the central motivation behind her project: she was determined to give other girls the opportunities she initially lacked—a chance to meet women pilots and get introduced to this exciting field. Her own path was tough enough, requiring immense dedication. She started her official flight training at just fifteen years old. Though her parents were incredibly supportive, she faced huge financial hurdles since her family "does not come from money".

The technical demands were also relentless. She needed to develop the "academic grit" to analyze complex concepts such as weather, engine mechanics, electrical systems, and navigation maps. This rigorous academic foundation proved that aviation was far more challenging than just flying a plane. By launching her project, Pepper vowed to simplify the entry point for girls, ensuring they had strong female role models to guide them through the complex and rewarding world of flight.

Building Workshops and Airfoils

To tackle the lack of exposure and the need for female role models, Pepper designed a massive, hands-on outreach project that she called "Bringing Aviation to My Community for Young Women". Her goal was to host comprehensive workshops in different areas across Southwest Florida— specifically Fort Myers, Naples, and Punta Gorda— to reach as many young women as possible.

These half-day sessions were a mix of science, engineering, and career counseling. Pepper, drawing on her own training, led many of the lessons herself. She taught the girls the fundamentals of flight, including the Bernoulli principle and how lift works. The curriculum was dense, at times challenging for younger participants (around age 10), but the hands-on activities kept everyone engaged.

The girls were given a rare chance to engage with true engineering. One of the main projects required them to use woodworking tools to build their own airfoils, which are the cross-sections of an aircraft wing. They also learned practical skills, such as flight planning, which involved reading navigational maps and calculating required fuel and distance.

To directly combat the lack of female mentors, Pepper made key partnerships:

Coordinating guest speakers from the Paradise Coast 99s chapter.

Inviting commercial and corporate pilots who fly for a career or recreation.

Connecting the young women to professional female pilots from the all-women pilot organization.

Working with local airports (FBOs) to secure necessary facilities.

This intensive preparation demanded over 120 recorded hours, much of which was spent on tedious but necessary tasks, such as "cutting and laminating" instructional materials. By ensuring that young women were taught by and worked alongside accomplished female pilots, Pepper achieved her goal of making the future of aviation visible and attainable for the next generation.

The High-Flying Hurdles

$ $ $ $ $

Pepper's path was fraught with challenges that tested her resilience, including the immense financial burden of learning to fly. Her private pilot's license alone cost around $16,000. While she demonstrated incredible resourcefulness in securing a $10,000 scholarship from the Aircraft Owners and Pilots Association (AOPA) and additional funds from the Jack Kent Cook Youth Foundation, the constant need for funding added pressure to her already demanding schedule. She remarked that applying for and securing scholarships had become routine for her.

The technical demands of pilot training brought personal setbacks that required her to overcome self-doubt. She vividly recalls the difficulty of mastering the "flare," the critical maneuver needed for a safe landing. This difficulty led to her biggest challenge: being so nervous that she failed her first practical exam. Having to retake a portion of the test was difficult, but it taught her a crucial lesson in pushing past failure.

The entire process was squeezed into her already packed high school schedule, which included demanding classes, cheerleading, lacrosse, and theater. The time commitment for her project easily surpassed the minimum of 80 hours, requiring her to dedicate over 120 recorded hours to the work. She had to deal with the exhausting logistics of preparation, which included securing facilities and gathering supplies, a process that was often "tedious".

To make this balancing act possible, Pepper made a key decision to switch to dual enrollment, taking college classes while still in high school. This adjustment provided her with the schedule flexibility needed to complete her rigorous flight training and execute her project successfully without compromising her education. Her perseverance during this time was later demonstrated when she earned her Instrument Flight Rating (IFR)—a significantly more difficult certification—while also working as a camp counselor for an aviation program.

Wings of Change: Inspiring the Next Generation

The true victory of Pepper was not just in completing her planned workshops, but in hosting the "fly day" where the young women could finally experience aviation for themselves. About 30 girls attended this climactic event, soaring through the sky in aircraft piloted by women mentors and other supportive pilots.

The impact was immediate and visible: many participants were instantly hooked, showing their immense enthusiasm. In a sign of lasting impact, several girls were so inspired that they registered for the IARA program to obtain their student pilot certificates, marking the official first step toward their own flight training.

However, Pepper knew the local success needed a national voice. She recognized that for young girls across the country to easily find aviation resources through Girl Scouts, the organization itself needed to change. Therefore, she drafted an advocacy letter and formal proposal to the Girl Scouts of the USA (GSUSA). Her petition requested that GSUSA "reinstate an aviation program or badge" that had been cut years earlier due to complex legal and liability concerns.

Pepper strongly advocated that aviation deserved recognition and support alongside other STEM and business programs that GSUSA promotes. This direct approach paid off handsomely: GSUSA invited her to meet with their programming and curriculum council, showing that they were taking her work seriously. While the official badge creation process is long (GSUSA estimated a potential release between 2025 and 2030), her efforts had an immediate ripple effect. Local councils were inspired by her advocacy, leading them to create their own patches and events, such as the aviation day hosted by the Sanimco council in Houston. This powerful advocacy proved that a local project driven by passion could lead to national policy conversations and inspire real change.

Cleared for Takeoff: A Future of Flight

Completing her demanding project taught Pepper invaluable lessons that extended far beyond the mechanics of flight. She learned that she possessed the "perseverance and like motivation" necessary to pursue ambitious goals. Overcoming major setbacks, like failing her first practical exam, instilled a profound sense of self-confidence that she carries into her young adulthood.

Today, Pepper is a college student at Rice University in Houston, Texas, where she majors in engineering and explores astrophysics. Her experience in balancing high school and her project prepared her to manage a full college schedule. She has deliberately sought a wider range of experiences, realizing that she regretted focusing so narrowly on engineering and aviation for so long.

She is now committed to "sample everything," balancing her technical curriculum with classes like philosophy and psychology, finding new interests in the study of the brain. This open-minded approach ensures she continues to grow and learn.

Pepper's future is rooted in the legacy of her project: she aims to continue inspiring others with her expertise. She hopes to become an engineer who also teaches people how to fly as a side business, utilizing her certified flight instructor rating. This goal to mentor and educate others is the enduring proof of her project's impact, guaranteeing that her passion for aviation will continue to propel young women toward the skies.

Chapter 3
Engineering for Real Life

Carson Kosar (Ep 29)

For Carson Kosar, the world of service wasn't just about local bake sales or park cleanups; it was a journey that unexpectedly took her thousands of miles away. While her primary academic interest lay in hospitality and tourism management—a field focused on service and planning—a pivotal community trip in 2017 revealed a problem that required a completely different set of skills: engineering and international logistics.

That year, Carson traveled to the Dominican Republic (DR) with a community service company called Fathom. Her goal was simple: to help wherever she could. In a neighborhood marked by poverty—a "not a great part of town" by her description—Carson met the incredible women of Ray Pel, an all-woman business specializing in creating beautiful products from recycled paper. Despite the challenges surrounding them, the women were full of infectious joy, character, and spirit, even singing as Carson's group approached. Their resilience immediately captivated her.

The women of the all-female business Ray Pel in the Dominican Republic create and sell recycled paper goods. As the women demonstrated the intricate process of turning waste into paper, Carson noticed a profound weakness in their system. At one stage, they needed to smooth the newly formed sheets of paper. Instead of a modern, efficient tool, the women used old roller-ball deodorant containers. This manual technique was not only "very time-consuming and hard" but also made the entire process difficult for the women who relied on this business for their livelihoods.

The image of that inadequate tool stuck with Carson long after she returned home. She was entering high school and actively searching for a project idea, something impactful that wasn't "too small". The memory of the women and their need transformed from a simple observation into a mission. She decided that her project would be to design and build a durable, custom-made stainless-steel tool—something resembling a professional paint roller—that would streamline the paper-smoothing process for Ray Pel. This commitment forced her to step entirely outside her comfort zone and into the challenging "realm" of engineering and manufacturing, a world she knew "had no idea what I'm doing in that realm" about.

The Forge of Fabrication

Carson knew that having a passionate idea was only the beginning; making a stainless-steel tool would require skills and connections she did not possess. To navigate this technical challenge, she tapped into her personal network. Her grandmother, who worked at a company connected to many engineers, became her critical link, setting her up with two different fabricators who could help bring the product "into fruition".

The design itself was in Carson's head, and she tried to sketch it out for the professionals, doing her "best I can to want I want to know what this product looks like". Communication issues and delays immediately plagued the work. The first fabricator lived an hour away, and after their initial meeting, contact stalled. Carson realized that, while she was "counting on him," she was not his "top priority." The project ground to a frustrating "standstill" while she waited.

Determined not to let the project die, she found a second, more responsive fabricator. This contact quickly provided prototypes, but Carson faced disappointment: the first two versions "didn't work" when tested against the actual paper samples from Ray Pel. This forced her to confront failure directly, and she learned that the path to a finished product is rarely straight. She was deeply committed to making a difference, recognizing that despite the setbacks, going from seeing the project "in my head" to "really coming to life" was the goal.

To execute the core of her project, Carson relied on a strong combination of persistence, teamwork, and strategic outreach:

Fabricator Network: She sourced and managed relationships with two separate metal fabricators to ensure a successful tool was built.

International Contact: She maintained contact with the women of Ray Pel for a year before the delivery, ensuring they wanted and needed the assistance.

Travel Logistics: She meticulously planned a return trip to the Dominican Republic to deliver the new tools and personally train the women.

Multifaceted Communication: She overcame the language barrier by using both translators and Google Translate to bond with the women during the hands-on training sessions.

It was only after the second prototype failed that the original, slower fabricator called back, having completed his version of the stainless-steel tool. This final, third version proved to be the correct solution, a powerful reminder that persistence pays off, especially when facing repeated rejection.

The Hardest "No" and the Real-World Test

The fabrication process wasn't the only challenge Carson had to overcome; managing her demanding schedule and navigating cross-cultural communication were equally challenging. Carson was a varsity athlete, held a job, and, most remarkably, was completing her associate's degree in hospitality and tourism management before she had even received her high school diploma. When asked how she handled it all, she emphasized that prioritization was key: "You're always going to make time for the things... that you think are important". This meant learning to manage her time effectively, overcoming her tendency to "procrastinate like crazy", a skill she refined throughout the demanding process.

About a year after she developed the tool, Carson returned to the Dominican Republic to deliver the metal-smoothing device. The setting was different—the group of women was smaller, and they had moved houses—but their joy remained. Carson was careful not to assume the women would automatically appreciate or use her creation; she emphasized that she would never "force anything on them", demonstrating a respect for their autonomy and culture.

The training itself was a logistical challenge due to the massive language barrier. Carson had studied only a single semester of Spanish, while the women primarily spoke Spanish or Creole. She relied heavily on translators and technology, using Google Translate to facilitate conversations and bond with the women. This commitment allowed her to spend genuine one-on-one time with them, learning about their lives, their families, and sharing meals. This intensely personal experience solidified her connection to the project, teaching her that overcoming the difficulty of communication was just as important as perfecting the stainless-steel tool itself.

The Triumph of the Failed Attempts

The final three days spent at Ray Pel were filled with profound moments of impact and unexpected learning. When Carson introduced the successful stainless steel smoothing tool, the women were immediately delighted, incorporating it into their routine and starting to make "a lot of paper" efficiently. Carson felt the intense satisfaction of seeing her vision directly improve their business operations.

But the most moving part of the trip was what happened next. Carson had brought along the first tool made by the second fabricator—the prototype she considered one of her "failed attempts". She had packed it just in case they could find some secondary use for it. The women, with their deep knowledge of the paper-making craft, realized that the size and shape of the "failed" tool were perfect for the paper-drying stage of the process.

This moment became Carson's defining takeaway, a powerful metaphor for the entire journey. She realized that what she viewed as a technical failure was, in fact, an "amazing success" when viewed through the eyes of the people it was meant to serve. This unforeseen repurposing taught her that effort is never truly wasted. She had successfully stepped far outside the comfort zone of her hospitality studies, delving into engineering, global travel, and cross-cultural communication, all of which culminated in a massive "ripple effect"—a core concept she champions in her advocacy. The project's impact was visible not just in the quantity of paper produced, but in the improved quality of life for the talented women of Ray Pel.

From Service to Future Vision

The dedication Carson showed to the women of Ray Pel was simply an extension of the discipline she applied to her own packed life. Managing a project that required overseas travel and technical collaboration, all while finishing her associate's degree three weeks before high school graduation, was only possible due to her fiercely honed skills in organization and prioritization. She learned that even when motivation lagged, the structural support of her team—her parents, friends, and mentors—was essential for pushing her forward.

Carson's project not only produced a tangible product but also yielded immense personal growth. It cemented her understanding that she is "capable of doing amazing things" and proved that if she puts her mind to something, she can accomplish it. This self-assurance directly influenced her public advocacy.

She took the central lesson of her project—that "stepping outside of your comfort zone can create a ripple effect"—and turned it into her message for a TEDx presentation. This was a long-held goal, and she used the platform to share how her uncomfortable dive into engineering changed her life and the lives of the women in the DR. This commitment to service also translated into political interest; she became an honorary legislative page, running errands and sitting in on the Senate and House of Representatives in South Carolina.

Looking ahead, Carson is using the foundation of her associate's degree to launch a career as an event planner in hospitality and tourism. She planned to use a gap year to secure local internships to further her career foundation before continuing her education. Her entire journey, from the poverty-stricken neighborhood of Ray Pel to the TEDx stage, proved that she possesses the confidence and organizational structure necessary to pursue any plan she sets her mind to. Her biggest advice to younger girls facing a similar project remains: "You are the only person who is setting your own limits."

Chapter 4
The Power of Robotics

Madalyn Nguyen (Ep 65)

The Personal Mission

Madalyn Nguyen knew precisely where she wanted to focus her monumental project: the world of science, technology, engineering, and mathematics (STEM). Her inspiration wasn't just a fleeting interest; it was a deeply personal passion sparked back in the sixth grade when she first began working with robotics. In that field, she quickly realized she was often one of the few girls in the room, which made her determined to pursue a future as an engineer. This dedication led her to notice a bigger problem in the world: the vast gender gap in STEM. For Madalyn, the project became the perfect tool to tackle this significant issue, transforming her goal into "bridging the STEM gender gap". She was primarily focused on making science and robotics accessible to girls in underrepresented and underserved communities. This was not a quick decision for Madalyn; her dedication to service and STEM had been building since she was young, beginning with her earlier projects, as both her Bronze and Silver Awards focused on STEM and engineering themes. Madalyn's commitment allowed her to achieve a major feat: she completed her entire project exceptionally early, wrapping it up by the end of her sophomore year. This massive undertaking required her to put in around 419 hours of work. She later shared that, even though the number

—

sounded huge, the time never truly felt grueling because she was genuinely passionate about her mission, believing that focus and determination are the keys to overcoming inevitable roadblocks.

Action at Home and in the Inner City

Madalyn designed her project to be multifaceted, ensuring she served her local San Diego community while also expanding her reach globally. Locally, she knew that young girls needed hands-on experience and visible role models. She partnered with her high school's robotics team and the local Society of Women Engineers (SWE) Next clubs to put on several events aimed at inspiration. One successful local event she organized was the "Girls in Inspiration Day," which impacted over 100 girls and their parents. This event gave attendees the chance to work on various engineering projects and participate in hands-on experiments, offering an authentic taste of engineering concepts. She also provided critical mentorship and guidance by organizing multiple First Lego League Junior Expos and actual competitions for younger students interested in robotics.

However, one of the most heartwarming local successes was her work at Wilson Middle School. This school served a low-income, inner-city community with limited resources, making it difficult for students to explore costly programs like robotics. Madalyn felt a personal connection to the school since her mother was an alumna. She successfully launched the school's first-ever First Robotics program. Despite initial challenges with resources and student recruitment, the program found its footing and began to flourish. Within two and a half years, the robotics program had tripled in size, and impressively, female students made up one-third of the participants. Today, that program operates completely self-sustaining, meaning the current students handle their own fundraising and management, securing a lasting educational opportunity for future generations in that community.

Bridging the World

Madalyn's dedication extended far beyond her San Diego neighborhood, encompassing a global vision to promote STEM education. Her international involvement began unexpectedly through social media. While scrolling, Madalyn and her mother discovered a First Tech Challenge (FTC) team in Texas, Team Mario, that was mentoring a First Lego League (FLL) team in Paraguay. Madalyn reached out and offered her experienced mentorship, and what followed was a life-changing trip. In November 2019, during Thanksgiving break, Madalyn traveled to Paraguay. During her almost week-long stay, she worked directly with the FLL team, going beyond virtual support to host an in-person First Tech Challenge scrimmage. This gave the Paraguayan team crucial face-to-face interaction and guidance. The mission was so successful that Madalyn began working to establish a Society of Women Engineers (SWE) club in Paraguay, demonstrating her ongoing commitment to creating sustainable resources wherever she went.

Madalyn's project implementation required immense coordination and leadership:

Political Advocacy: She traveled to Sacramento with her mother to discuss STEM-related bills and rights with legislators.
International Mentorship: She collaborated with an FLL team in Paraguay and a high school team in Texas.
Community Building: She secured space and participants for large-scale local events, such as the Girls in Inspiration Day (100+ attendees).
Curriculum Development: She engineered and managed the creation of the first-ever First Robotics program at Wilson Middle School.

Madalyn noted that a significant challenge during this international phase was the language barrier; she had only a couple of years of high school Spanish, and the students in Paraguay spoke "very, very little English". However, she remained determined, focusing on the higher goal of empowering the students, which allowed her to overcome the communication gaps.

Overcoming Roadblocks and Finding Confidence

💪💪💪💪💪

Everyone pursuing a project at this level encounters challenges, but Madalyn faced unique hurdles both personally and logistically. She started the award very early in her high school career—as an underclassman—which meant that she "was a little bit more timid" and faced practical limitations. For example, she couldn't drive, so she often relied on her mother, who became her essential "Uber everywhere". Her early start also meant she had to learn to lead and organize large events and competitions with diverse groups of people, including adults and students who spoke little English, which she admitted was "a little bit difficult". She overcame these issues by focusing on sheer determination. She said there are "no like instructions, guidelines to doing any of this. You kind of just got to jump head in and just be confident and just try your best". This perseverance helped her grow immensely, helping her "break out of my shell a little bit to make me the person I am right now".

The most significant strategic pivot came unexpectedly with the COVID-19 pandemic. Although Madalyn had been lucky enough to finish most of her major projects before the widespread lockdowns, she realized the project shouldn't just be something "that you do for the award". When in-person activities stopped, Madalyn strategically used the available technology, realizing she could turn the challenge into a "good thing". By hosting virtual events using Zoom and social media, she could reach a "much wider audience and honestly make a bigger impact than I was making in person". This virtual expansion allowed her to connect with robotics teams and students globally, including in Mexico and Paraguay, demonstrating how quickly she could adapt her original vision to achieve broader, sustained impact.

Signature Moments and the Future

🚀🚀🚀🚀🚀

Madalyn's hard work generated a measurable and visible impact. Her commitment and expertise led to her being selected as a keynote speaker for a local SWE (Society of Women Engineers) event in San Diego, where she spoke to over 400 people about mentorship, role models, and overcoming hardships in the male-dominated STEM world. She also presented at the international SWE conference.

However, the most significant impact was felt on a personal level. Madalyn shared that her favorite memory highlights the true success of her role as a mentor: she was working with a First Lego League team composed mainly of girls when they approached her and asked for her signature on their posters. Madalyn found this funny, admitting she wasn't a celebrity, but realizing that these girls looked up to her as a role model. Looking to the future, Madalyn is focused on academic excellence. It has been her lifelong goal, since she was eight years old, to attend the California Institute of Technology (Caltech). She plans to major in engineering, possibly pursuing software engineering. Madalyn advises future candidates to be highly determined, suggesting they start their project early to navigate challenges and give themselves extra time. She also strongly recommends using social media, calling her project's dedicated Instagram account one of her "best resources" for documenting her work and networking.

Madalyn's journey from a timid middle school student to an internationally recognized STEM leader illustrates that the project is not merely a destination, but a launchpad; it's the moment she realized the full horsepower of her passion, shifting from simply admiring role models to stepping into the spotlight and becoming the one who signs the autographs for the generation that follows.

Chapter 5
Exploring Engineering Careers

Evelyn Taliaferro (Ep 105)

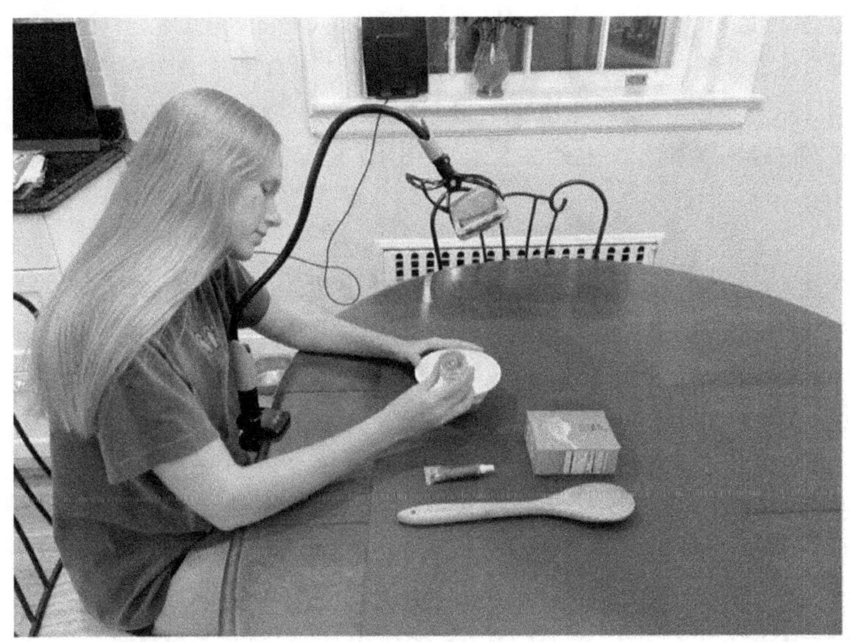

Evelyn Taliaferro spent her middle school years like many students: unsure about the path her future might take. When people asked what she wanted to study or what career she dreamed of pursuing, she felt like she had "absolutely no idea". It wasn't until she reached high school that her future finally clicked into place.

Evelyn decided, somewhat on a whim, to apply for a specialized science program called Project Lead the Way. She was just looking for something interesting to try. But once she started taking those introductory classes, everything changed. She found herself hooked, realizing, "Whoa, this is actually so interesting". That program gave her the exposure she needed to fall in love with engineering.

This sudden discovery of her passion led to a deeper realization: why hadn't she known about engineering sooner? She realized that if she had received that exposure sooner, she might have started thinking about a career in STEM (Science, Technology, Engineering, and Math) much earlier. But what if she hadn't randomly applied to that high school program? She might be doing something "completely different with my life" right now.

This became the "main driving force" behind her project. Evelyn recognized a problem in her community: young girls, especially those in elementary and middle school, weren't being exposed to the incredible world of engineering during those crucial developmental years. She knew that early exposure was key, regardless of gender, to show kids all the exciting options available to them. Evelyn decided her mission would be to throw open the doors to engineering, making sure other girls wouldn't have to stumble into it by chance as she did. She was determined to create a project that would ignite that spark of interest in them, giving them the confidence to pursue whatever they were passionate about.

The Crash Landing and the New Flight Plan

When Evelyn first began her journey, she was running blind. Her troop hadn't had anyone complete the award before, so she had no one to show her the way. She focused on a topic she was only "mildly interested" in, and she submitted her first project proposal. The response was brutal: her proposal was "completely rejected" by the council. It wasn't just a few tweaks she needed; the council essentially told her to scrap her entire idea and "start completely over from scratch".

This moment felt "demoralizing and a little bit like frustrating and disappointing" because she had already invested so much time into writing that first proposal. Dealing with such a huge failure right at the beginning of the biggest project of her life was difficult. She had to step back and decide whether to let this stop her.

Looking back, however, Evelyn learned that the rejection turned out to be the "best thing the council could have possibly done" for her. The original project wasn't STEM-related, and she wasn't genuinely passionate about solving that specific issue in her community. The rejection forced her to rethink everything and go back to her deepest passion: engineering.

The biggest shift she made was mental. Instead of choosing a project she thought sounded interesting and then trying to find a problem it solved, she flipped the process around. She realized she needed to focus on the community need first—the lack of exposure to engineering—and then design a project that leveraged her strengths and passions to address it. This new approach strengthened the second proposal, made it more focused, and, most importantly, made it more meaningful to her. She was now working on an issue she deeply cared about, making the long process ahead much more rewarding. Evelyn learned a critical lesson right away: sometimes, a failure is just a redirection to a much better path.

Building a Blueprint for Discovery

Once her new project was approved, Evelyn launched her ambitious plan to bring engineering education directly to young students. Her goal was to make complex fields accessible and fun. She structured her project around two main parts: a comprehensive digital resource and hands-on community workshops.

The central piece was a detailed website that mapped out the thrilling world of engineering. Evelyn focused on the six major branches of engineering, offering clear explanations of what each branch is and the kinds of careers one could pursue within it. To make the careers real and relatable, she included valuable:

Interviews with female engineers working in those exact fields give the girls an inside look at their day-to-day lives.

Support from her mentor, Steve Webb, an engineer who personally reached out to his colleagues, persuading them to share their professional stories for the interviews.

Simple, easy-to-do DIY crafts and challenges tied to each branch, using items kids could find at home or easily buy.

The second major part was getting out into her community. Evelyn was determined that the project couldn't just exist online. She successfully partnered with local non-profit organizations to run hands-on workshops.

- She hosted two educational sessions at local nonprofits, Four Kids and the YMCA.
- She prepared a short presentation on the different engineering branches, brought all the supplies, and guided the kids through her favorite crafts.
- She partnered with a local K–12 school, giving them a large box of supplies and her website information so their elementary and middle school students could continue exploring the projects.

Her two favorite crafts clearly demonstrated the fun she was having as she brought the science to life. As an electrical engineering student herself, she loved the challenge of having students use wires, pennies, a nail, and lemons to build a lemon battery. But the kids' overwhelming favorite was the bottle rocket, an exciting explosion made with baking soda and vinegar inside a plastic bottle. This mix of digital resources and exciting hands-on fun ensured that students would not only learn the information but also remember how exciting science could be.

Rockets, Lemons, and Real-Life Role Models

The true success of the project wasn't just in the beautifully built website or the detailed lesson plans; it was in the workshop sessions, where Evelyn saw her vision come to life. Although she admitted she felt "very nervous before the workshops" about whether the kids would enjoy the material, their excitement was instantly rewarding.

The atmosphere at the workshops was electric. Evelyn saw many children start "a little bit confused and uninterested," especially when faced with the initial concept of "engineering". But once she guided them into the hands-on challenges and projects, they became completely absorbed. She watched them race to see who could build the strongest clay boat or cheered as they competed to create the "biggest bottle rocket explosion". This immediate, visible enjoyment was incredibly motivating for Evelyn, showing her that all her research and hard work were paying off.

Beyond the explosive fun, Evelyn knew the power of seeing successful women in these roles. Thanks to her mentor, Steve Webb, the website provided direct access to female engineers who shared their stories. These interviews acted as real-life proof that engineering was a career path where girls could succeed and thrive.

Evelyn's efforts created a powerful, ongoing wave of impact. The partnerships she forged guaranteed that the lessons would continue even after she moved on. The materials she donated to the local K–12 school would benefit future students through her curriculum. Her project proved that exposing young people to STEM through play and real-life inspiration could break down intimidating barriers, one lemon battery and one soaring bottle rocket at a time. This visible excitement from the young students was the "incredibly rewarding" feeling that fueled her to the finish line.

Becoming Part of a Global Sisterhood

Completing the project wasn't the end of Evelyn's journey in engineering advocacy; it was only the beginning of her leadership growth. The confidence she gained from leading workshops and recovering from her initial setback carried her forward, even as she started her college career at Princeton University.

Evelyn quickly sought out opportunities to continue the work she loved. She stumbled upon the Society of Women Engineers (SWE) on campus, an organization that immediately resonated with her mission. She learned that SWE hosts large workshops for local girls, bringing them onto campus to hear presentations on engineering, which is what Evelyn's project was about.

"When they told me about that, my jaw just dropped because I was like This is exactly what I have done over the past year with my project. So, this would be perfect," she recalled.

She immediately dove in, helping to organize these massive events. Her first workshop with SWE hosted over a hundred girls, a huge step up from the groups of 20 to 30 she managed for her project. She found the experience just as rewarding as her original project, confirming that spreading excitement about engineering was her lifelong passion. Now, she is dedicated to becoming more involved, having applied to oversee the SWE's girl program for the following year.

As she pursues her major in electrical and computer engineering, Evelyn has embraced the open-minded attitude that the project taught her. She knows that even though she started "running blind" and faced massive rejection, she has the resilience to take on any challenge and the wisdom to know when to ask for help. Her journey proves that leadership means being flexible and knowing that even failures can lead to meaningful successes.

Chapter 6
Apps That Make Life Better

Amelia Dunkin (Ep 145)

Amelia Dunkin grew up in a town she thought she knew well. As a self-proclaimed "huge history nerd," she loved diving into the past. She even considered majoring in history when she headed off to college. Yet, like many people who live in the same place their whole lives, she was completely unaware that her own city, Lamont, was full of hidden secrets and surprising historical events.

Her world changed when she finally secured a spot on one of the Lamont Historical Society's walking tours. These tours were hard to join because they were held only annually or semiannually, and space was limited to just 20 people at a time. She discovered that even though she had lived in Lamont her whole life—a town her mother had also lived in for years—she had "never heard of any of this" incredible local history.

The most exciting revelation centered on the local stone Metro station, a small building she had probably walked past hundreds of times. She learned that this seemingly ordinary building had witnessed key moments in American history, including the passing of President Lincoln's funeral train. Even more dramatically, it was a major site in the historic Lamont massacre. Townspeople gathered there hoping to corner a garrison of soldiers sent from Chicago to stop a quarry workers' strike, although the soldiers ultimately got off at a different stop.

This experience was Amelia's "main driving force" for her project. She realized that if she, a passionate history lover who lived in the town, struggled to learn these facts, everyone else must be missing out, too. She saw a clear problem: the existing tours were not accessible to the public due to limited availability, time constraints (two hours), and the physical demands of walking up and down Lamont's famously hilly terrain. Amelia was determined to share these "hidden gems" of history with everyone in her community, turning her passion for the past into a digital blueprint for the future.

From Spiral Notebook to Swift Code

Amelia decided the best way to make her town's history available to everyone, regardless of schedule or mobility, was to create a location-aware walking tour application (app). This was an ambitious choice, especially since she planned to do the bulk of the intense work under a tight deadline.

Her first hurdle was the technology itself. Although she already had coding experience, she was building an app for Apple's iOS system, which meant she had to learn an entirely different language to program it. Amelia knew how to code small programs, but transitioning to a visual, platform-specific language was a major challenge. Learning this new language took a significant amount of time and became one of the most time-consuming parts of her entire project.

Next came the equipment barrier. To create an Apple app, she needed specific, specialized software called Xcode. The tricky part? You can only run Xcode on an Apple computer. This put a significant dent in her plans, as she realized she might need to spend a lot of time fundraising just to afford the hardware. Fortunately, Amelia found a quick solution: her older brother let her borrow the computer that he was using for college, avoiding the need for a massive fundraising campaign just to buy a machine.

Despite the complexity of app development, Amelia recognized that addressing technical challenges was necessary for her project's long-term impact. The result—an app that guides users by showing their location and giving directions to the following historical site—would ensure people wouldn't get lost while learning. Her dedication to mastering the new programming language and navigating the specific hardware requirements showed that she was willing to push her technical skills far beyond what was easy to achieve her goal.

Mapping the Hidden Gems

The power of Amelia's project came not just from the app's code, but from the network of local connections she built to gather and deliver the historical content. She understood that she couldn't create a comprehensive historical resource on her own, so she leveraged her community's expertise. She worked with groups ranging from history buffs to artists and even her high school speech team.

First, she collaborated directly with the Lamont Historical Society, the very organization that had sparked her idea. They provided her with a crucial starting point: a list of the sites they already used on their limited physical tour, along with research materials. She also talked to her local history teacher, who provided a list of "a few other sites" he used when teaching his students, ensuring her tour went beyond the standard narrative.

To turn these notes into engaging content, she knew she needed more than just historical facts. She needed fresh voices and creative elements. Amelia reached out to student groups at her high school to gather resources:

High School Historical Honor Society: Students helped write the site descriptions.

—

Art Honor Society: Designed the app's logo, which was important since Amelia admitted she "cannot do art or graphic design".

Speech Team: Helped record the audio descriptions for the app.

Lamont SESA Centennial Committee: Provided her with an event platform to promote the app during Lamont's 150th anniversary celebration.

These partnerships helped Amelia maximize her resources, logging an estimated 100 hours of volunteer time (not including her own programming hours). The finished app offered users a choice of three different tours, including a full 31-site tour, a shorter tour for those with mobility issues or small children, and a specialized tour focusing on the many historic churches that dot Lamont's skyline. By including the audio descriptions recorded by the speech team, the app allowed visitors to hear the history explained, making the experience dynamic and accessible even while walking.

The Trials of Apple and the Clock

If learning a new programming language was Amelia's biggest self-challenge, then battling time and the official approval systems were her major external roadblocks. The entire approval process was incredibly time-sensitive, and she quickly learned that external organizations rarely move on a student's deadline.

Amelia began the formal process a year before her due date, yet it took the committee "like six months to actually get it approved". This slow timeline was incredibly stressful because she still had to teach herself the necessary programming language in the remaining few months. She realized that to manage the paperwork efficiently, she should have written much longer, more detailed answers in her initial proposal so she wouldn't have to keep submitting revisions for review.

Once the content was ready, the biggest challenge was navigating the rigorous process of getting the app onto the Apple App Store. Apple has a "very interesting screening process," and only a small percentage of apps submitted are approved. Amelia's app failed the first time because it was location-aware. Apple's developer testers were in California and too far from the Lamont area to view the content so that the app wouldn't display correctly for them. Amelia had to figure out this obscure technical error and then ask the testers if they could use a VPN (Virtual Private Network) to simulate being in Lamont so they could successfully test the app.

Another major unanticipated hurdle involved legal forms. Because the app "does track location," Amelia suddenly had to create legal documents, a task she had not been expecting to do as a high school student. These obstacles—from waiting for the council's slow approval to fighting the App Store's technical screening and drafting legal paperwork—taught her quickly that adapting to setbacks and staying persistent were more important than sticking to an initial plan. She needed patience when things did not follow her ideal timeline.

Beyond the Code: A New Path Forward

Amelia's finished app had an immediate, measurable impact on her community. The app launched just in time to be featured as part of Lamont's 150th anniversary celebration. She printed 500 business cards with a QR code linking directly to the app store. She gave half of them to the Historical Society to ensure the tour remained accessible and sustainable long after she finished her project. By putting the app in the free Apple App Store, she ensured the tour could reach anyone interested, whenever they wished. The Historical Society can now give the cards to visitors who want to download the tour themselves.

The actual growth, however, was in Amelia herself. Before starting the project, she confessed that standing in front of people and talking was "not an option". Even taking a speech class was difficult. The app development process forced her to interact with teachers, historians, town committees, and even Apple App Store developers. By the end, she realized she had gained several crucial skills that she now takes into her future, including learning how to write a professional email and the confidence to stand in front of groups.

Amelia is now attending the University of Chicago and plans to major in computer science and history (or economics). Her experience of blending history and technology has inspired her career goals: she hopes to work in machine learning and Artificial Intelligence (AI). She specifically wants to combine AI with history to "search through primary sources" or to create better economic models in the future. She realized that her project showed her that she has the drive to achieve challenging goals.

Chapter 7
Building Coding Confidence

Kennedy Watkins (Ep 102)

For Kennedy Watkins, the incredible journey toward her project began with a simple question she asked herself when she was just nine years old: *What if I could build that?*

Kennedy's love for technology and science started early when she attended her first coding camp. Like many kids discovering a new passion, the experience felt magical, a feeling reinforced when she successfully built her first application. This game lets users try out different virtual nail polish colors on their fingernails. This early success gave her a rush, proving that coding wasn't just complicated math; it was a creative superpower. As she grew older, this initial spark turned into a serious passion that she desperately wanted to share with others.

In high school, Kennedy joined a unique extracurricular program dedicated to technology and education called "one laptop per child". The students in the class spent their time learning how to use a simple programming language called Scratch, which uses easy-to-understand "block code". Together, they created their own lesson plans, learning how to break down complex instructions into simple steps for beginners.

The high point of this class came when Kennedy and her classmates got the chance to turn their hard work into real-world action: a trip to Barbados. Traveling overseas for the first time, Kennedy helped teach Scratch to elementary students there and shared their newly developed lesson plans with the local teachers. This experience was more than just a cool trip; it was a fantastic lesson in how young people can use their knowledge to help others. When she returned home to the United States, Kennedy knew exactly what her next big mission had to be.

Bringing the World Home

Returning to the Columbus City School District after her eye-opening trip, Kennedy saw a clear opportunity. She realized that if she could share her love of coding in Barbados, she could certainly share it with the students in her own city. She decided her project, which she named "Let's Code Together," would be dedicated to bringing the excitement of STEM back to the elementary and middle schoolers right in her area.

"Let's Code Together" was designed to address the lack of early exposure to computer science. Kennedy's core belief was simple: the sooner kids learned that coding was a viable, fun career path, the better. She wanted to show them that science careers weren't limited to laboratories; they also involved creative skills, like using Scratch to build games and solve problems. She planned a two-part approach:

First, she would create a digital learning resource. This involved building a comprehensive website to host all her materials, including detailed online lesson plans and easy-to-follow tutorial videos. She wanted this digital platform to be a permanent resource that students and teachers could use anytime, anywhere, ensuring the project would continue long after she was gone.

Second, she planned a series of hands-on, interactive workshops. The original goal was to mirror the successful in-person teaching model she had used overseas. She aimed to partner with at least three local classrooms, going in person, giving PowerPoint presentations, and guiding the young students through the lessons. Each session would not only teach them basic coding but also introduce them to various STEM careers, inspiring them to ask questions like, "What does a coder do?" and "How can I be a coder now?" By the end of this planning phase, Kennedy felt confident. She had the perfect model, a specific goal, and a deep, personal passion for the subject.

The Unexpected Roadblock

Kennedy's carefully laid plans hit a wall: the COVID-19 pandemic.

The bulk of the work on her project suddenly had to be done while the world was shutting down. Her entire original plan—visiting classrooms, meeting students in person, and running highly interactive sessions just like she did in Barbados—was now impossible. Instead of face-to-face teaching, she had to switch to Zoom.

This change created a massive challenge that threatened to sink her project: getting into the classrooms at all. Teachers were already overwhelmed. They were struggling to balance hybrid learning, managing some students in person while others were online, making their schedules "very hectic". They were incredibly reluctant to interrupt their precious hour of class time just to let a high school student come in and take over, even for a valuable coding lesson.

Kennedy realized she couldn't overcome this challenge alone. She was trying to break into a stressed, locked-down system, and her typical approach of tackling everything by herself wasn't going to work.

Luckily, Kennedy had a powerful resource: a dedicated mentor, a retired teacher and family friend. This mentor understood the challenges teachers were facing and knew how to navigate the school system.

Kennedy's mentor connected with teachers and persuaded them to incorporate "Let's Code Together" into their virtual classrooms. This crucial support enabled Kennedy to overcome resistance and secure the classroom time needed. It was a big step out of her comfort zone, forcing her to rely on others' wisdom and connections. She learned a vital lesson about leadership during this difficult time: true success isn't about doing everything yourself, but about trusting others and delegating tasks effectively.

Let's Code Together in Action

Once the virtual workshops were running on Zoom, Kennedy finally saw the direct impact of her hard work. She worked primarily with students in grades 4 and 5. The structure of her one-hour virtual classes was carefully organized: she introduced herself, led the main coding lesson using the block-based programming language Scratch, and then dedicated time to discussing different STEM career paths.

The students' feedback was overwhelmingly positive. They were thrilled with the experience, especially excited about the games they were able to create in such a short time using a platform they didn't even know existed. The students were eager to show off their creations and were instantly curious about careers in the field. When they asked how they could become coders, Kennedy gave them the best possible news: "Congratulations, you're all coders." Seeing the smiles on their faces after this announcement was a fantastic moment for Kennedy.

To execute and sustain the quality of her project, Kennedy relied heavily on the team she built:

A **retired teacher mentor** who helped her gain essential classroom connections.
Supportive teachers who became impromptu team members by recommending her program to their colleagues.
An **editor** who helped create, polish, and manage the tutorial videos for the project website.
Her **mother**, who suggested sharing the project outcomes on social media.

This digital outreach proved far more powerful than Kennedy initially imagined. After her mother suggested uploading information about "Let's Code Together" to social media, teachers from entirely different states began contacting her. They reached out via direct messages (DMs) requesting the lesson plans and files, demonstrating that Kennedy's influence extended far beyond city limits, impacting children across the nation. This realization gave her a "new perspective on how big my impact was and how far it could last".

From Code to Agriculture

The complex process of managing her project forced Kennedy to recognize her own strengths and weaknesses, leading to significant personal growth. She realized that before this project, she preferred working alone, relying solely on her own capabilities. The project was a powerful lesson in the value of communication and teamwork, teaching her that it was okay to delegate tasks and trust that her team would complete them to her standard.

This newfound entrepreneurial confidence and platform-building skills led Kennedy to launch a new business, Gilded Teas. Gilded Teas is a loose-leaf tea company where Kennedy creates and sells her own unique tea blends. However, her commitment to STEM and her project legacy remains central to her business model. 10% of all Gilded Teas profits are donated to organizations that help children learn to code, such as STEM camps and "hackathons" that first inspired her when she was young.

Kennedy's motivation for her tea business came from her family's non-profit garden club project, where she developed an interest in growing herbs and spices. This passion has now grown into a large-scale future goal: she is currently working

—

towards creating a tea farm in Florida in the coming years, where she hopes to grow all her own ingredients. As she plans her future in agriculture, she has realized a new need: she wants to use her business platform to support organizations that help minorities and women enter the agriculture field. In this sector, both groups are underrepresented.

Kennedy's entire journey, from overcoming rejection to building a lasting national impact through both technology and entrepreneurship, demonstrates a deep commitment to service. Her project proved that she could transform passion into a powerful, permanent force for change.

Chapter 8
Robots, Coding, and Cookies

Makayla Hoefs (Ep 153)

Makayla Hoefs' journey, titled "Coding for Cookies," didn't begin with a typical service project; it started in the high-tech world of her school's robotics team. About two and a half years before her project became official, Makayla and a teammate who was passionate about getting girls involved in robotics noticed a gap. While Girl Scouts is well-known for cookies and camps, Makayla realized that many younger girls were missing out on the thrill of science, technology, engineering, and math (STEM) fields, especially robotics.

The very first session they hosted was a huge hit. They invited a local troop to their robotics facility and showed them the ropes. The students learned how to program a small Lego robot using code and even got to take the controls of an FTC robot (the kind used in competitive high school robotics) to learn how to drive it. They also introduced the girls to their high school's big competition robot, showing them just how far they could take their interest. The response was immediate and overwhelmingly positive: the girls "really loved it and enjoyed it".

Over the course of a year, Makayla and her team ran these sessions for troops in their local area, specifically in Becker and Big Lake, Minnesota. This initial effort taught over 100 girls about robotics. However, the program eventually reached a turning point. They had successfully reached most of the local troops, and Makayla wanted to take "Coding for Cookies" further, expanding its reach far beyond her own town. That desire to create a long-lasting, widespread impact is what made the program the perfect foundation for her project. Her goal was to make robotics and programming accessible not just to a few local troops, but to girls across the state and beyond, using her passion for STEM to inspire the next generation of female coders.

The Challenge of Statewide Scale

Once Makayla decided to turn the successful local program into a project, she faced a significant hurdle: how to serve a huge geographical area with limited resources. Her council, Minnesota Wisconsin Lakes and Pines, covers a vast region with many small towns spread far apart. If she wanted to reach more girls, she couldn't rely on troops driving hours to her local facility. She needed to make "Coding for Cookies" mobile and accessible to anyone in the council area.

The solution required a massive collaborative effort with her Girl Scout Council. The council itself already ran a system called "program on the go," in which staff members would travel long distances to deliver kits and activities to those located far from headquarters. Makayla decided to integrate her robotics program into this existing successful structure.

To do this, she needed two things: official council approval for the sessions and, crucially, a way for the council to run the sessions *without* her robotics team present. The council agreed to help with promotion and registration, making the "Coding for Cookies" sessions official council-sponsored events publicized on their website and in program guides. This single action instantly opened the doors across the entire state.

However, the bigger challenge was creating a program so simple and well-documented that any adult—even one without robotics experience—could pick it up and teach it. Makayla and her robotics team had to design a teaching tool that could operate perfectly every time and a manual that explained complex robotics in the simplest terms. This was difficult because Makayla, who prefers to handle tasks herself, had to learn to trust her teammates with crucial aspects like writing and programming to ensure the program was easily replicable and sustainable for years to come.

The Nuts and Bolts of Sustainability

The core of Makayla's project, and the key to solving the challenge of scalability, was the creation of portable robotics kits. These kits transformed the robotics lab's specialized equipment into a simple, ready-to-use learning platform that the council could send to distant troops. This idea perfectly fulfilled the sustainability requirement, ensuring the program would thrive long after Makayla finished high school.

The complexity lay in designing kits that could be used by anyone, regardless of their technical background. Makayla's robotics team pitched in, with certain members focused on their strengths. They accomplished this goal through several key actions:

Pre-Programming Robots: They programmed the small FTC robots so they could be turned on and used immediately by non-technical staff.

Creating a Comprehensive Manual: They wrote a detailed, step-by-step instruction manual to accompany the kit, ensuring that any troop leader or council member could easily teach the session.

Aligning with Girl Scout Goals: They created a binder that linked the program's activities directly to the badge requirements for different age levels, making it clear how troops could earn badges by participating in the program.

Designing Reusable Materials: The kit contained all necessary materials for a full "Coding for Cookies" session, making it a complete, self-contained educational tool.

Makayla successfully convinced her robotics team to build and donate a complete kit to the council. The council instantly embraced the concept, turning the kit into an official "program on the go" resource that staff could travel with to teach girls across the region. Furthermore, the robotics team kept additional kits on hand to lend to other robotics clubs, allowing the program to reach even more communities outside the direct council network. By making the instruction manual and a project presentation available online, Makayla provided a template that other robotics teams anywhere could follow to replicate her program.

Measuring Success in Confidence and Code

Makayla and her team carefully tracked the success of their sessions by gathering specific data from the participants. After each "Coding for Cookies" event, attendees were given an evaluation form to complete. The results proved that the hands-on approach was not only fun but highly effective at boosting confidence and interest in STEM.

The measurements showed that the girls involved achieved significant learning outcomes:

- They all gained a "basic understanding of first robotics," the competitive world in which Makayla's team operated.
- Every girl learned how to code a Lego robot, drive an FTC robot, and operate an FRC robot.
- They rated their overall session experience with high marks; the lowest score was 6 out of 10, and the majority gave 9 out of 10.

Perhaps the most impactful result was the change in how the girls viewed their own abilities. The program received a 97% positive leadership experience rating, indicating that the participants felt more confident about pursuing STEM fields after attending. The data showed that the girls particularly enjoyed the practical, hands-on learning, confirming Makayla's belief that early exposure through engagement is the best way to ignite a passion for technology. The net promoter score—a measure of how likely the participant is to recommend the program to a friend—was an incredible **77%**, demonstrating that the girls were leaving the sessions not only informed, but genuinely excited to share their new coding knowledge.

The culminating personal memory for Makayla came after she presented her final report via Zoom to the review committee. When she walked out of the room, her entire robotics team was waiting, armed with gold balloons, ready to celebrate her success. This heartwarming moment symbolized the actual achievement of her project: bridging the gap between her two most important communities, robotics and Girl Scouts, in a fun and meaningful way.

Delegation: The True Leadership Upgrade

The project proved to be a powerful learning experience that transformed Makayla's approach to leadership and teamwork. She confessed that one of her biggest personal struggles was her tendency to take on everything herself, preferring to get tasks "done like myself". The time-consuming nature of creating the kits and writing the complex, instructional manual forced her to rely on others.

She learned that asking for help and delegating weren't signs of weakness but signs of effective leadership. By letting go of control, she utilized the specific talents of her robotics teammates—the members who were excellent at programming handled the robot setup. The members who were better writers managed the manual. This experience of working through her proposal revisions taught her the value of researching thoroughly and continually refining her work, two skills that are essential far beyond her project.

Makayla's journey emphasized the power of perseverance. Even though she faced a lengthy initial proposal process, going back and forth with revisions, she didn't get discouraged because she understood that the revisions were meant to make her project perfect and ensure its long-term success. Now that she has seen her idea through to completion and created a lasting resource for her council, Makayla encourages other girls to pursue their ideas. She sees the project not as a daunting task, but as a chance to see a deeply cared-for idea realized.

Looking ahead, Makayla's future is rooted in the success of her project. She continues to be a leader, acting as a resource and providing support to younger girls working on their own projects. By tackling the most challenging part—the planning, research, and delegation—Makayla secured a future for "Coding for Cookies," inspiring countless girls who might otherwise never have touched a robot.

Chapter 9
New View on Ability

Stella Kaval (Ep 48)

The Foreign Word

For a long time, Stella Kaval felt a strange sort of hesitation whenever she encountered people with physical disabilities. It wasn't that she didn't want to be kind; it was simply that she didn't know how to act or what to say. She had grown up in a world where disability inclusion and accessibility weren't parts of her formal education, leaving a gap in her understanding of others. One term felt especially mysterious and out of reach: cerebral palsy. To her, "the word cerebral palsy was foreign," and she realized that this lack of knowledge was likely shared by many of her peers.

Stella understood that cerebral palsy, or CP, is a condition that affects the way a person moves their muscles. It is the most common motor disability found in childhood, affecting roughly 17 million people across the globe. Despite how common it is, Stella noticed that people with the condition were often treated as if they were fundamentally different from everyone else. She wanted to prove to her community that just because someone is physically disabled, "it doesn't mean they aren't just like you and me".

This realization sparked a massive undertaking. Stella decided her project would focus on changing the perception of disability in her community

—

through education and technology. She didn't want to give just a single speech or hang a poster; she wanted to create tools that would last and help bridge the social gap between people with disabilities and their typical peers.

Before she could teach others, however, Stella knew she had to become an expert herself. She began a deep dive into the world of CP, conducting intensive research and meeting with community members to hear their stories firsthand. This wasn't just library work; it was a journey into the lives of doctors, engineers, and families who dealt with these challenges every single day. Each interview added a new layer to her understanding and fueled her drive to make a real difference.

Listening To The Heart

One of the most pivotal moments in Stella's research was her meeting with author Becky Taylor. Reading Taylor's book, *Tell Me the Number Before Infinity*, opened Stella's eyes to the specific barriers children with CP face in school settings. She learned that inclusion isn't just about being in the same room; it's about feeling like you belong and having the same opportunities as everyone else. This conversation shifted Stella's focus toward the importance of empathy in the classroom.

Stella also interviewed Alva Gardner from Ability Now Bay Area. Alva shared the frustrating reality of the lack of accommodations at the university level and how those missing resources can derail a student's entire education. Today, Alva spends her time helping adults with disabilities reach their full potential, and her wisdom helped Stella see the long-term impact of her project. It wasn't just about helping kids; it was about preparing them for a life of success and independence.

The emotional core of the project truly came together when Stella met Jacquie Robison, the founder of the nonprofit WAWOS. Jacquie shared the story of her daughter, Sofia, and described how isolating it can be to live with CP when the world isn't built for you. Jacquie pointed out a significant problem that Stella felt she could address: a severe lack of activities to help children learn about empathy and disability. Hearing about Sofia's experiences firsthand made the mission personal.

Armed with these stories and scientific data, Stella identified her target audience as youth ages 8 to 15. She believed this age group was at the perfect stage to learn about inclusion so they could actively include other kids on the playground and in their classes. She wanted to take the isolation Jacquie had described and replace it with a community where every girl and boy felt seen and understood.

Building A Bridge With Technology

▦ ▦ ▦ ▦ ▦

To turn her research into action, Stella set out to create two resources: a hands-on activity pack for classrooms and a mobile app called "Ability All". The activity pack was designed to be "kid-friendly" and easy for educators or troop leaders to use, featuring 9 activities that teach empathy and disability inclusion. Stella had a blast with the graphic design part of the booklet, playing with color palettes and inserting images to make the material engaging for younger students.

The creation of the "Ability All" app was a much bigger technical challenge. While Stella had taken some basic computer science classes in high school, she had never been asked to look at technology through the lens of disability. She had to completely rethink how a person with limited motor skills would interact with a screen. This meant her coding process had to be precise and thoughtful to ensure the app was truly accessible to everyone.

To make "Ability All" a reality, Stella followed a detailed technical and organizational process:

Accessibility Integration: She researched and implemented features such as alternative text for images, large click targets for more

Motor Skill Development: She designed specific games within the app to help children with cerebral palsy practice their motor skills while interacting with their families.

Peer Mentorship: She recruited other girls in middle school to join her team, training them to lead the nine activities in the booklet for elementary school students.

Outreach Strategy: She established a weekly schedule for her team to visit local schools and youth groups to implement the curriculum.

Realizing that technology could be used as a force for good in the accessibility space changed Stella's entire perspective on her future. She learned that coding isn't just about moving a motor or making a game; it's about solving real-world problems for people like Sofia. Through extensive editing, testing, and hard work, she proved she could create high-tech solutions that benefited people of all abilities.

Rolling With The Punches

Just as Stella's project was gaining momentum, the world changed. The pandemic hit, throwing a massive wrench into her carefully laid plans. One of the biggest disappointments was the cancellation of an event she had organized at the Oakland Athletic Stadium. She had already recruited and organized a large team of volunteers to teach youth about inclusion through hands-on activities at the stadium, but the event's safety was suddenly compromised.

Instead of giving up, Stella showed the resilience of a true leader. She knew that the need for disability education didn't go away just because people were stuck at home. She quickly pivoted, moving all her resources and activities into a digital format. She created a project website where the activity pack could be downloaded and encouraged her volunteers to spread the word online. This shift allowed her message to reach people who might not have been able to travel to the stadium.
https://www.changingtheperceptionofdisability.org/

There were other, more personal challenges as well. When Stella went into classrooms to teach younger children about disability, she sometimes encountered students who asked difficult questions or even joked about the topic. It was frustrating and sometimes hurtful to hear, but Stella handled it with grace and respect. She realized that these moments were exactly why her project was so necessary.

She learned to lead these tough conversations by staying calm and helping the children understand what it means to have a disability. When she saw the "lightbulb moment" in a child's eyes—when they finally understood that their peers with CP were just kids who wanted to play—she knew she was making a lasting difference. Overcoming these social and global obstacles only strengthened her confidence and commitment to the cause.

A Legacy Of Inclusion

The impact of Stella's project was visible in the smiles of the people she served. One of her favorite memories was seeing Sofia and other children with cerebral palsy using the "Ability All" app. Seeing the hard work she put into the accessibility features and user interface pay off was incredibly inspiring. It wasn't just a school assignment anymore; it was a tool that was helping families connect and children grow.

Stella also saw the fruits of her labor through the many speeches she delivered to young girls and students. After one presentation, a student came up to her and said, "I didn't know what cerebral palsy was, but now I do and now I know to include everyone on the playground". These small interactions proved that the perception of disability in her community was truly shifting. Stella had successfully empowered a new generation to lead with empathy and kindness.

Beyond her project, Stella's journey as a girl had been filled with many other adventures. As a younger Girl Scout, she loved camping at Camp Skylark and Camp Bothin, where she had her first experiences away from home. In 8th grade, she traveled to London to see the global impact of her organization, and in high school, she served as a national delegate. She was eventually appointed to the board to help shape policies for thousands of other girls. These roles taught her how to speak up and represent others on a large scale.

Looking toward the future, Stella has set her sights on a career as an engineer. She wants to continue working in accessibility, developing new technologies that help people with disabilities navigate the world more easily. Her project didn't just help her community; it helped her find her own purpose in life.

Chapter 10
STEM for Social Good

Supriya Kotnani (Ep 110)

From a young age, Supriya Kotnani knew she wanted a career dedicated to helping people heal. Driven by her passion for health and science, she set her sights on becoming a physician. Before heading off to college, she gained hands-on experience by interning at a local medical clinic. It was here, surrounded by doctors and patients, that she found the powerful spark that would ignite her project, "Heartbeat for Healthy".

During her internship, Supriya noticed a frustrating pattern. Patients would come in with health concerns, and the doctor would often give the same simple advice: "Oh, eat healthier". This instruction sounded great, but Supriya quickly realized that for many people, simply knowing they *should* eat better wasn't enough. The real problem was that they didn't know **how** to eat healthier. She watched patients walk in, sometimes literally carrying food from the Taco Bell just a short distance away. She realized that the barrier wasn't desire; it was accessibility, affordability, and practical knowledge. How could people choose a healthy, homemade meal when they didn't know how to make one quickly and affordably, especially when a fast-food burrito was so close and easy?

Supriya felt she had identified a genuine need in her community: bridging the gap between telling people to eat healthy and providing the practical steps to do so.

Her first idea was a familiar solution: pamphlets.
She thought producing bright, informative
pamphlets filled with health tips would be a genius
idea, a solution worthy of a Nobel Prize. However,
her liaison quickly pointed out a fatal flaw. The
liaison asked Supriya how many times she had
picked up a pamphlet and used the information in
her daily life. The answer was clear: the pamphlets
weren't sustainable and wouldn't lead to a real,
lasting change in people's habits. This rejection of
her first plan was a tough hurdle, forcing Supriya to
restart her thinking completely. But by focusing on
the community's need first—the need for
affordable, practical recipes—she finally landed on
the perfect approach for her project.

The Recipe for Success

After overcoming the initial setback of her rejected
pamphlet idea, Supriya designed a new strategy
that was practical, affordable, and, most
importantly, delicious. She recognized that any
successful program needed to reach the people
who needed it most: those living in food deserts. A
food desert is an area where access to fresh,
healthy, and affordable food is limited, often
surrounded instead by fast-food restaurants.

Supriya and her team worked tirelessly to identify
these areas. She created a large spreadsheet
listing all the neighborhoods that lacked grocery

stores and were surrounded by fast-food chains. Then, she reached out to residential neighborhoods and childcare facilities in those areas, offering to partner with them to bring "Heartbeat for Healthy" directly to their residents.

The key to her new approach was the interactive nutrition sessions. She crafted these sessions to be immediately engaging and understandable. The educational part included a detailed PowerPoint presentation that didn't just lecture; it clearly listed the food's health benefits. Her goal was to "entice them," letting them know, "Look, it's great for your brain. It's great for your heart". This was immediately followed by the most crucial step: a taste test of the healthy recipe. The taste test proved that healthy food could be genuinely good.

To ensure the lessons were sustainable, Supriya focused heavily on making the recipes affordable and repeatable. Participants received special recipe cards that included step-by-step instructions and, critically, a list of prices. Her organization maintained a strict budget rule: the subtotal for all ingredients had to be less than $10.

These foundational elements created a project that truly impacted daily life:

Identifying Community Need: Using demographic data to target food desert locations.
Affordable Meal Planning: Ensuring the total cost of ingredients was under $10.

Interactive Learning: Combining visual presentations with real-life taste testing.

Supriya wanted participants to be able to make a direct comparison: "This is what a typical burrito at Taco Bell costs versus this very healthy, nutritious, affordable recipe that you can make in your everyday household". This simple comparison empowered people to choose wellness for themselves and their families.

Convincing the Community

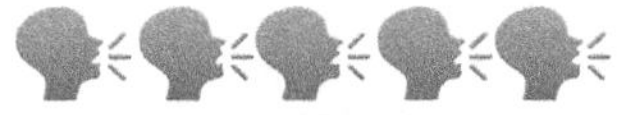

The path to turning "Heartbeat for Healthy" into a widespread success was not easy. Supriya faced resistance not only when her initial pamphlet idea was shot down but also when she attempted to expand her program into new geographical areas.

After launching successfully in her home community in California, Supriya moved away for college, attending Baylor University in Texas. She immediately saw a need to continue her mission in her new environment, the Waco community. However, convincing a whole new group of organizations and people that a new student's project was worth their time proved challenging. Supriya found that people often wouldn't take her seriously because they didn't know who she was. She had to essentially "convince the Waco community that we are interested in helping them".

This hurdle forced her to rely heavily on the leadership skills she had developed in Girl Scouts, which taught her the "value of teamwork", foundational confidence, and the importance of having a reliable support system. Her organization thrived because she wasn't alone. She had a strong core team, including Darlene Chen and Sophia, who shared her mission and worked with her daily. She joined around seventh grade, specifically looking for an organization that focused on community service. Her troop leader, Miss Kelly Black, encouraged her and helped her build the confidence to overcome challenges, such as public speaking.

Supriya learned that perseverance was necessary to build credibility from scratch. She knew she had to continually reach out, present her ideas clearly, and trust that if she kept working hard, people would eventually take her seriously. She demonstrated resilience, knowing that the project's success depended on her ability to turn small wins—like getting one or two partners interested— into a strong community resume that convinced others to join her cause. The lesson she took away was that using her voice to make a change was paramount, even when it felt like she was at the bottom of the ladder.

The Taste Test and Traveling Impact

The impact of "Heartbeat for Healthy" was immediately measurable through the positive feedback Supriya received from participants and partners. After holding over 30 nutrition sessions in both California and Texas, Supriya consistently reported that "most people really, really like the recipes that we make." The hands-on taste testing sessions worked perfectly, proving that healthy food was palatable and desirable.

Beyond taste, the sessions provided powerful education. Supriya's curriculum focused on translating nutritional facts into easy-to-understand health benefits, emphasizing why certain foods were "great for your brain" and "great for your heart". This educational component ensured that participants understood the *why* behind the nutritional advice, encouraging them to try making the recipes at home.

The ultimate measure of success for "Heartbeat for Healthy" was its lasting reach and its ability to replicate its mission in new communities. Supriya successfully expanded the program from her hometown to the community around Baylor University, demonstrating that her model was strong enough to convince new partners in a different state. This expansion proved that her foundational work wasn't just a temporary project but a sustainable blueprint for change.

The entire process had a profound effect on Supriya's personal growth, particularly in developing her confidence and public speaking skills. She noted that managing relationships with her team members, mentors, and community partners forced her to master effective **communication**. Before the project, she preferred working alone, relying only on her own skills, but the project taught her that "it was okay to rely on someone else" and to delegate tasks to her talented team. This shift from working alone to leading a collaborative team was essential to the project's ability to maximize its impact in two states. The positive change she created in the community, simply by sharing knowledge and delicious, low-cost recipes, was powerful enough to motivate her long-term goals.

From Heartbeat to Healing

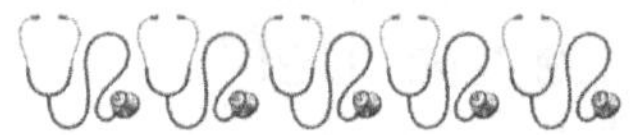

Supriya's experience with the project solidified her career aspirations and provided her with confidence and invaluable professional skills. As she pursues her degree on the pre-med track, she is determined to become a physician. Crucially, she plans to use the insights she gained from "Heartbeat for Healthy" to shape her future practice. Seeing patients struggle with nutrition-based concerns firsthand taught her that health isn't just about medicine; it's about addressing basic needs like diet. The project taught her that she must use her "voice to make a change" and help alleviate problems, even if she can't solve every single one.

Her focus on nutrition and leadership extended beyond the project itself. She continues to lead "Heartbeat for Healthy," planning to expand it further in California and Texas, with a long-term aspiration to take the concept global.

Supriya credits the award requirements for providing the structure for her project and for building her personal confidence. She learned how to understand her weaknesses to build a strong team to turn them into strengths. This realization that leadership requires relying on others has been key to her success.

She encourages girls to find a project they are truly passionate about because when the inevitable challenges arise, that passion will be the fuel needed to "see it through". Supriya's journey proves her project is a springboard for lifelong commitment to service and leadership.

Supriya's Heartbeat for Healthy project is like an unexpected seed dropped into dry soil: though the first attempt (the pamphlet) failed, the second attempt focused on providing nourishment (affordable recipes and education) to fill a fundamental deficiency, allowing a lasting, vital plant to grow, helping whole communities bloom with better health and knowledge.

Chapter 11
Bringing Tech to New Places

Mehaa Amirthalingam (Ep 109)

Mehaa Amirthalingam discovered her mission halfway across the world. In 2019, she visited Champa, Cambodia, and traveled to a small village called Maliki. What she found there wasn't just a distant community, but a place struggling with a huge problem: a serious "education divide". The children living in Maliki Village were bright, sharp, and incredibly eager to learn, but they faced constant obstacles. They were isolated, cut off from the big urban and commercial areas, which made transportation impossible. Even worse, many families were locked into a cycle of poverty, meaning children—especially girls—were forced to skip school to work and earn money for their families.

For Mehaa, this wasn't just a sad statistic; it was a call to action. She saw how incredibly "capable" these children were, realizing that despite their isolation, they retained so much information. They just weren't being given the basic tools they needed to succeed. Mehaa explained that these children could break "away from the shackles of this poverty that they have been connected to for their entire lives", but they needed a hand—they just needed "like one push to have extra resources where they can move out of this place".

Her time in Maliki Village crystallized her purpose. Before long, she was thinking about her project, realizing that creating a computer lab in this village would be the perfect way to address the issue. The goal was huge: providing accessible education, especially for young women and children denied schooling. Mehaa decided that if she could make this happen in Maliki Village, her project could serve as a "blueprint for a solution" that other people could use to address the same problem in countless other poverty-stricken places around the globe. She was determined that this small, isolated village would become "exhibit A of what could be done" when dedicated people offer support.

Planning and the Pandemic Wall

Mehaa began turning her passion into a plan, knowing that her project needed to bridge that huge geographic and educational gap. She knew that providing physical computers was the first step, but she quickly discovered that securing the necessary technology wasn't as difficult as she thought. She simply started "talking to my parents and friends" in her small circle, and soon, people who weren't using their computers were willing to donate them. The initial donations focused only on computers that were running Windows 10 and MacBooks, ensuring the children received high-quality resources that would last for years.

The campaign grew like "a wildfire," spreading through friends, family, and even friends of friends.

Soon, emails and texts were flooding in from people all over, eager to donate resources. Mehaa noted that people were motivated because they recognized that the lack of education for poverty-stricken children was a global problem. Maliki Village was just a "small part of this big puzzle" that could be solved one piece at a time.

However, just as her plan was building momentum, the COVID-19 pandemic hit.

The pandemic completely changed everything. Mehaa had always envisioned her project as "so hands-on," with her surrounded by people. Now, all her outreach and teaching had to happen online. The core challenge became isolation. Mehaa admitted that much of her project, estimated at "80% of my project," felt lonely because she was conducting research and building the online platform entirely by herself. Another major unexpected hurdle was "How are we going to get the computers to Cambodia?" With worldwide travel blocks in place, there was massive uncertainty about when, or if, volunteers would be able to transport the collected computers safely. Despite the difficulty, Mehaa found that facing these issues head-on meant she had to grow in flexibility and perseverance.

The Worldwide Web of Help

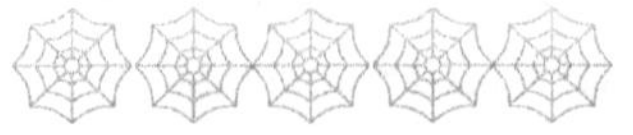

To make the computer lab a permanent reality, Mehaa needed strong partners. She connected with Maliki World Missions, the nonprofit organization that had initially built the community center where the lab would be housed. This collaboration was essential because the founder and director, Miss Leng Abbassi, provided vital "firsthand help," answering questions about the tiny, microscopic issues residents faced and managing the final logistical step: transporting the computers from Texas to Cambodia.

Mehaa realized that while she was the leader, she couldn't achieve this massive goal alone. She relied on a dedicated team for everything from transportation to tedious administrative tasks.

Mother: Drove Mehaa to downtown Houston and other locations to pick up donated computers.

Friends: Helped with both the research needed for the project and the technical effort of **refurbishing the computers**.

Volunteers (Maliki World Missions): Ultimately traveled to Cambodia to install the computer lab.

Advisor Miss Becky: Provided "specific feedback" on the reports, which Mehaa admitted were often frustrating and tedious due to the council's strict review process.

Since the project was conducted almost entirely during COVID-19, Mehaa focused on creating a robust digital classroom platform. She gathered online resources and uploaded them to the platform, ensuring that the Maliki Village students, despite their isolation, would receive "basically the same education as us," since everyone around the world was relying on online platforms at the time. She also structured the system to be sustainable. By partnering with Maliki World Missions, she ensured ongoing support. The organization sends monthly volunteers, and teachers are there who "are able to take care of the computer lab and make sure it's in good condition," so the children can continue to use the resources efficiently. The dedication of her team and the community proved that even in times of crisis, "people are still willing to help in whatever way they can".

From Frustration to Fulfilled Promise

Pushing through the isolation and the constant administrative hurdles of the proposal required tremendous mental strength. Mehaa faced times when she thought her project was "impossible" because of the massive uncertainty surrounding the pandemic. She spent hours alone, writing and researching, realizing that "95% or I guess 80% of my project was a bit isolating," which surprised her.

To keep herself going during the rough times, Mehaa relied on a specific, powerful memory: the children's faces. She had visited the community center before the pandemic and led a one-day workshop focused on arts, crafts, and English. This experience, where she was "immersed into a community," was the fuel that kept her working. She remembered watching the children, noting how their "eyes lit up at like flashcards" and how excited they were over simple items like Styrofoam letters. She realized then that if these children were so eager to learn from basic resources, "the world is their oyster" when given full access to 15 laptops and computers, on par with access in first-world countries. This memory became her "constant like fuel for like the gear to continue with my project".

Her initial workshop also provided a sneak peek into the learning they would do in the new lab. During the workshop, she exposed them to "computer literacy," teaching them what keys to use and how to navigate a device. This focus on building their "digital language" proved that even a small effort could instantly improve their relationship with education.

Once the computers were finally installed, the positive impact was enormous. Even just 10 to 15 laptops were "all it took for them to expand their knowledge and their education". Most importantly, the women and girls who had been forced out of school early were now able to gain access to a basic education. The community center became a vital space where children could gather after working their strict family hours, finally being able to "be children for once".

A Legacy of Light and Future Blueprint

The enduring success of Mehaa's project lies in its carefully planned sustainability. By donating the refurbished computers and establishing the online learning platform, the computer lab continues to be used "so efficiently". The partnership with Maliki World Missions ensures that the equipment remains maintained by monthly volunteers and dedicated teachers. This powerful, self-sustaining model means that the impact of her project is not confined to one year; it is a permanent educational lifeline for the village.

Mehaa realized that the lessons learned from the project went far beyond simply building a computer lab. She gained confidence in networking, encouraging others to "be involved with your community" and "don't be afraid to reach out" to organizations, knowing that nine out of ten people are eager to help. She learned that finding passion might happen unexpectedly—her focus on education equity was something she discovered right before her project, not something she had been planning since she was in fifth grade. The journey taught her that while the project requires immense work and can be difficult, having a supportive team and deeply believing in the mission are essential to seeing it through.

Today, Mehaa is studying mechanical engineering in college. Her ambition remains focused on service and global change. Although her project was about technology and education, she is also involved in environmentalism, having developed a "dual water toilet flushing system" to save freshwater, which is a major passion project. She loves "storytelling" through various outlets, including talking about her project and the global water crisis on platforms like Snapchat. Whether she is pursuing engineering or leading service initiatives, Mehaa's core goal remains the same: to continue giving back to the community and to be surrounded by others who share a passion for service. She reflects that the final project was the "cap off" to her Girl Scout career, fulfilling her desire to give back to the community that raised her.

Mehaa's journey, which overcame the challenge of distance and isolation by creating a permanent educational resource, is like building a network of satellite dishes: Though the receivers (the students) were in a remote location and the signal (the education) was initially blocked, she focused on technology and communication to boost the signal, creating a powerful, reliable connection to the entire world of learning.

Chapter 12
A Cosmic Mission

Gitika Gorthi (Ep 69)

The Spark of Ambition

For Gitika Gorthi, the night sky was never just a collection of distant lights; it was a roadmap for her future. From a very young age, she found herself caught between two fascinations that most people think are worlds apart: the infinite reaches of outer space and the intricate workings of human medicine. While many of her peers were focused on local hobbies, Gitika was looking toward the stars, wondering how the human body could survive there. This unique combination of interests eventually became the foundation for a mission that would take her from her own backyard to a global stage.

She realized early on that there was a gap in the way people viewed space exploration. Most students saw it as a field strictly for pilots or engineers, but Gitika saw it as a frontier for doctors and researchers. She wanted to show other students, especially girls, that the space industry was vast enough for every kind of passion. This realization wasn't just a passing thought; it was a call to action that led her to begin her highest and most ambitious project yet. She knew she wanted to create something that didn't just stay in her town but reached out to the entire world.

The inspiration for her initiative came from a desire to ignite a fire in others. She saw that many youths were interested in science but felt intimidated by the complexity of the space sector. Gitika decided that she would be the one to bridge that gap. Her project wasn't just about finishing a requirement for her youth organization; it was about building a community of "thinkers" who were brave enough to ask big questions about our place in the universe. She began her journey by researching the current state of space travel, including the shift toward privatization, and how these changes would create new opportunities for the next generation of leaders.

Setting out on this path required immense courage. Gitika had to balance her intense school schedule with her growing passion for space medicine. She didn't want just to be a student of the stars; she wanted to be an advocate for them. This meant stepping into the role of a leader long before she felt fully ready. However, she leaned into the idea that if she didn't take the first step, no one else would. She was determined to prove that a single girl with a vision could launch a global conversation about the future of humanity in space.

Into the Digital Frontier

Turning a dream into a global presence required Gitika to become a master of technology and networking. She knew that if she wanted to reach students in different countries, she couldn't rely on posters or local workshops alone. She needed a digital home for her mission. Gitika set out to build an international website that would serve as a hub for space education and expert interviews. This wasn't a simple task for a high school student; it involved learning the "ins and outs" of web design and content creation while ensuring the information was accessible to everyone.

The execution of her project was a lesson in meticulous planning. She had to identify partners who shared her vision and convince them to support her cause. Gitika didn't just wait for people to find her; she actively hunted for connections that could elevate her platform. She looked for doctors at major space agencies, researchers in space medicine, and leaders in the private space industry. Each email she sent was a tiny launch, and with each response, her network grew stronger and more diverse.

To manage the complex stages of her project, Gitika followed a particular flight plan for her work:

Platform Development: She designed and launched a professional website to host resources, articles, and a series of high-level interviews.

Expert Outreach: She contacted dozens of space professionals, utilizing a "nothing to lose" mentality to secure conversations with people she once only dreamed of meeting.

Multilingual Integration: She leveraged her multilingualism to ensure her message could cross borders and reach students who might not speak English.

Nonprofit Founding: She formalized her work by creating a nonprofit organization to monitor the website and ensure the mission would continue for years to come. https://www.ignitedthinkers.org/

As the website came together, Gitika realized the power of her own voice. She started conducting interviews that gave her audience an "inside look" at what it's really like to work in the space industry. She wasn't just a reporter; she was a student learning alongside her viewers. By creating this international presence, she proved that she wasn't just working on a project; she was running a professional organization. The digital frontier enabled her to overcome the limits of her age and location, allowing her to impact lives in countries thousands of miles away.

The Shot Not Taken

One of the most defining philosophies Gitika adopted during her project was the famous saying: "You miss 100% of the shots you don't take." This mindset became her shield when she faced the intimidating task of reaching out to world-class experts. It can be terrifying for a teenager to email a senior scientist at a space agency or a famous astronaut. Still, Gitika refused to let fear keep her on the sidelines. She understood that the worst thing that could happen was a "no," but a "yes" could change the entire direction of her mission.

This courage led to some of her favorite memories. There were people she dreamed of interviewing—professionals she viewed as heroes—who responded to her requests. When she finally sat down to speak with them, she realized that these leaders were excited to see a young girl taking such a strong interest in their field. These interviews weren't just about gathering facts; they were about building bridges between current legends and future pioneers. Gitika learned that persistence is the primary fuel for any achievement and that most doors will open if you are brave enough to knock on them multiple times.

The journey wasn't without its obstacles. Managing an international presence meant dealing with different time zones, technical glitches on the website, and the constant pressure to keep the content fresh and engaging. There were moments when the sheer scale of what she was trying to do felt overwhelming. She had to learn how to manage her time effectively, ensuring that her schoolwork didn't suffer while she was busy running a global nonprofit. She had to stay organized and disciplined, treating her project with the same level of seriousness that a professional would.

The biggest challenge was often internal. Like many girls taking on huge tasks, she had to battle the "what-ifs." What if the website crashed? What if no one watched the interviews? Gitika overcame these doubts by focusing on her "why." She remembered the student who might feel inspired by her work, or the girl who didn't know that space medicine was even an option for a career. By focusing on the impact she wanted to make rather than the perfection of the process, she was able to conquer her challenges and keep moving forward, proving that the only real limit was the one she set for herself.

A Multilingual Bridge

Gitika possessed a unique skill that set her project apart from many others: she was fluent in multiple languages. She realized that the stars don't belong to just one country, and the conversation about space exploration shouldn't either. By integrating her multilingual abilities into her project, she was able to offer a benefit to the space program that few others could. She used her languages as a bridge to connect with international students and professionals, making her website a truly global resource.

This aspect of her work highlighted the importance of inclusion and diversity in STEM. Gitika didn't want space to be a "closed club." She wanted a girl in India or a student in Europe to feel just as connected to her mission as someone in the United States. Speaking different languages allowed her to understand various cultural perspectives on space travel and medicine. It made her a better communicator and a more empathetic leader. She showed her community that being a well-rounded person—someone who values both science and the arts, like linguistics—is a significant advantage in the modern world.

The impact of this global reach was visible in the feedback she received. People from all over the world began visiting her site and engaging with her content. She was building an international presence that far exceeded her original expectations. Her website became a place where the privatization of space travel wasn't just a news headline, but a topic of discussion for the people who would be living through it. Gitika's ability to communicate across borders made her mission sustainable and relevant to a massive audience.

Through this process, Gitika learned that her identity was her greatest strength. Being a girl who loved space, a student who studied medicine, and a leader who spoke multiple languages made her the perfect person to lead this cosmic mission. She realized that the space program needs more than just technical skills; it needs people who can bring humanity together. Her project proved that when you use your unique talents to help others, you create a ripple effect that can span the entire globe—and maybe one day, the whole solar system.

Launching Into the Future

As Gitika looks toward the horizon, her goals are as big as the universe itself. Her experience running this project has cemented her desire to pursue a career in space exploration, specifically in space medicine. She has seen how her work can inspire others, and she is determined to keep leading the way. The project taught her that she can manage international-level logistics, found a nonprofit, and interview world-renowned experts. She is no longer just a student with a dream; she is a trailblazer with a proven track record of success.

Her advice to other girls is simple: find what they love and not be afraid to go "all in." Gitika's journey shows that you don't have to wait until you are an adult to make an impact. You can start right now, with an idea and the drive to make it happen. She believes that every girl has the potential to be a "changemaker" if she is willing to take the shot, even when it feels like a long way from the hoop.

Her leadership growth has been the most significant part of the journey. She evolved from a girl who was "foreign" to the complexities of the space industry to a confident advocate who can speak intelligently about the privatization of space and the future of medicine. She learned that being a leader doesn't mean having all the answers; it means being the one who is willing to ask the questions and find the people who can help answer them. The support of her team, her family, and her mentors was essential, but it was Gitika's own persistence that kept the mission on course. Her story is a reminder that while the stars may be far away, the path to reaching them begins with a single, courageous decision to lead here on Earth.

Chapter 13
Create Your Own
STEM Project

The Strength In Redirection

In the engineering design cycle, your first plan is often a "crash landing". It is common to start "running blind" and submit a proposal that is "completely rejected". While this moment can feel demoralizing and frustrating, it is a vital part of the testing process. A rejection is often a redirection to a "much better path" where you find your true passion rather than just a topic you are "mildly interested" in. For example, an initial idea to create educational pamphlets might be shot down because pamphlets aren't sustainable and rarely lead to real change. This failure forces the innovator to rethink and focus on what the community needs, such as interactive workshops or affordable recipes.

Embracing failure as the "best thing" that can happen allows you to flip your process and build a stronger, more focused "flight plan". You must first focus on the community's needs, then design a project that addresses them. This phase of the cycle is about testing the "sustainability" of your vision before you even build it. It requires the "academic grit" to analyze your own work and the resilience to take a step back when things don't work. Whether you are facing a "brutal" response from a review council or realizing your initial solution has a "fatal flaw," the key is not to let the setback stop you. Authentic leadership means being flexible and knowing that even the most significant failures can lead to the most meaningful successes. By treating every "no" as a data point, you move closer to a solution that will work in the real world.

The Loop Of Innovation

🔧🔧🔧🔧🔧

Once you move from planning to building, the "Test and Improve" phase becomes a hands-on battle with the "loop of innovation". Prototyping is the process of bringing an idea "into fruition" by creating early versions of a tool or app. It is common for the first few versions to "not work" when tested against real-world samples. For instance, a metal tool designed to smooth paper might fail twice before a third version proves to be the "correct solution". This process teaches you that the path to a finished product is rarely straight and that persistence is your most important tool. Even digital solutions face these tests; the App Store might reject an app for "obscure technical errors" or for location-aware settings that don't work for testers in different states.

One of the most surprising parts of testing is discovering that a "failed attempt" can still be an "amazing success". You might bring a rejected prototype along "just in case" and find that users find the size and shape perfect for a different part of the process, such as paper drying. This proves that effort is never truly wasted. Testing also means preparing for "unanticipated hurdles," such as legal requirements for location tracking, and "hectic" schedule changes due to global events like a pandemic. You must "think on your feet" and adjust your "workshop schedules" or teaching methods to meet the current reality. Some innovators even face personal setbacks, such as being so nervous they fail a practical exam, but they use the experience to build "perseverance and motivation". By gathering feedback and "measuring success in confidence," you ensure your final product isn't just a gadget, but a tool that changes how people view their own abilities.

Your STEM Solution Toolkit

Every big project starts with one small choice and a single moment of curiosity. This toolkit provides a collection of steps to help you launch your own journey into innovation and problem-solving:

• **Start by Noticing**: Look around your neighborhood, school, or park and ask, "What feels unhealthy, unsafe, or unbalanced?" STEM innovators look for "clues" or "signals" in their community, such as a lack of healthy food options in specific neighborhoods or students without access to technology. Identify a "fundamental deficiency" that needs a solution.

• **Learn Before You Act**: Use library books, science websites, and conversations with experts to research your solution before jumping in. This often requires developing "academic grit" to master complex subjects such as engine mechanics, navigation, or new programming languages. Knowledge is your strongest tool for avoiding accidental harm.

• **Build a Support Team**: No innovator works alone; you must build a team of classmates, teachers, and community leaders. Reach out to professionals for firsthand help, as adults are often eager to assist young leaders. A mentor, such as a retired teacher or an engineer, can provide critical connections to help you navigate school systems or technical hurdles.

• **Start Small, Stay Consistent**: Every massive project begins with a single idea and one person willing to try. Small steps stack up quickly, so choose one action you can take this week, such as organizing a single community event or creating one tutorial video. Consistency in your efforts is what eventually makes your impact "unstoppable".

• **Design for Sustainability**: Create a plan that allows your project to thrive long after you are finished. This can include building portable kits or writing a comprehensive manual so that anyone can teach the program without you being present.

• **Speak Up for Science**: Use your voice to share what you care about through presentations, posters, or videos. You can create national change by writing formal proposals or advocacy letters to large organizations, requesting that they reinstate programs or update their curriculum to include new STEM fields.

• **Celebrate Every Win**: Don't wait for a giant victory to feel successful; celebrate every person inspired and every small step forward. The actual achievement is becoming the kind of person who pays attention and acts. When you see your vision improve the life of even one person, it creates a "ripple effect" of positive change.

Launching a STEM project is like building a lighthouse: you must first find the rocky shore where help is needed, gather a crew to help you stack the stones of research and code, and keep the light of your passion burning through the storms of failure, knowing that your work will guide a whole new generation toward their own bright horizons.

ABOUT THE AUTHOR

Sheryl M. Robinson is a podcaster, mentor, and speaker dedicated to helping teens and young adults discover their unique gifts, talents, and abilities, creating a path toward their dreams.

Sheryl holds a Master of Arts in Servant Leadership from Viterbo University and a Bachelor's in Accounting from Southern Illinois University – Carbondale. She has been a proud member of Girl Scouts for more than 30 years. Her passion for supporting teens — especially those pursuing the Girl Scout Gold Award — led her to create *Hearts of Gold*, a YouTube series and podcast featuring Gold Award Girl Scouts from across the world.

In recognition of her work elevating and supporting the Girl Scout Highest Awards, Sheryl has been honored with the GSUSA Thanks II Award, the organization's highest recognition for service.

Recognizing the need for younger Girl Scouts to have resources and role models as they pursue the Bronze Award and Silver Award, Sheryl created this middle-grade book series to share inspiring stories of leadership, courage, and community change.

She deeply believes that the Girl Scout Highest Awards not only make the world a better place but also transform the Girl Scouts who earn them — building lifelong changemakers, confident problem-solvers, and compassionate leaders.

ACKNOWLEDGMENTS

Creating this book has been a journey shaped by many remarkable people, and I am deeply grateful for each of you.

To **my mom, Jean**, who first started me in Girl Scouts many years ago and planted the seeds of everything that would follow.

To **my daughter, Nikki**, a Bronze, Silver, and Gold Award Girl Scout whose dedication inspires me every day. Watching you flourish through each phase of your life is one of my greatest joys.

To **my husband, Mark,** thank you for always supporting me and the many plates you quietly set beside me while I typed away. I thank God for bringing you into my life every day.

To **Kamryn**, thank you for reading the first draft and sharing thoughtful feedback. Your insights helped shape this book and made it stronger.

To **all the Gold Award Girl Scouts** who have shared their stories on the Hearts of Gold podcast — thank you for trusting me with your journeys. Your courage, creativity, and leadership inspire thousands.

To the **Girl Scout leaders, volunteers, and parents** who support these incredible young women: your encouragement makes meaningful change possible.

To **Cassie**, who encouraged me to restart my Girl Scout journey when my daughter joined Girl Scouts.

A heartfelt thank you to **Stacie and Shannan**, who have listened to me talk about this book for years and never stopped encouraging me to make it happen.

To **Walter**, my podcast editor for the first nine years, and **Tommy**, my new editor — and to their entire family, especially **Greg**, whose podcasting challenge a decade ago helped set all of this into motion.

And finally, to **Elsie, Rob, Cliff, Daniel, and Jessica** — thank you for keeping the process fun, for sharing your knowledge, and for helping Hearts of Gold continue to grow.

This project exists because of each of you.
Thank you for helping bring these stories to life.

MORE STORIES

Want to hear more inspiring stories from Gold Award Girl Scouts?

HeartsofGoldPodcast.com

You can watch or listen to new episodes every month.

Podcast:
https://bit.ly/3JT7x0w

YouTube:
https://bit.ly/3P5nns8

Instagram:
https://bit.ly/3JZ2JX8